Preface

Welcome to the One For All Olympiads Science Class 8 Book!

This book is specially crafted to ignite the spark of curiosity and foster a love for science in young minds. Designed with the unique needs of first graders in mind, it provides a comprehensive introduction to fundamental scientific concepts through engaging, age-appropriate content.

Our primary goal is to make learning science an enjoyable and enriching experience. The book is filled with colorful illustrations, real-life examples, and interactive exercises that help children understand and relate to the world around them. Each chapter is carefully structured to build on prior knowledge, ensuring a steady progression in learning.

Key features of this book include:

Interactive Exercises: Hands-on activities and simple experiments designed to engage students actively and make learning fun.

Multiple Choice Questions: End-of-chapter MCQs help reinforce learning and provide ample practice for young learners.

Cross-Curricular Connections: Integrates elements of math and language arts, promoting a holistic learning experience.

Environmental Awareness: Introduction to basic environmental science and sustainability concepts.

Through this book, we aim to encourage critical thinking, problem-solving skills, and a deeper understanding of scientific principles. The inclusion of real-life scientists and their contributions will inspire young learners and highlight the importance of science in our daily lives.

Whether used in the classroom or at home, this book is an invaluable resource for fostering a lifelong love of science. We hope that it will be a stepping stone for students, helping them explore and appreciate the wonders of the scientific world.

Thank you for choosing our Science Olympiad Book. Let's embark on this fascinating journey of discovery and learning together!

Happy Learning!

Table of Contents

Questions: Part 1

1. What are the main components of food?

   a) Carbohydrates, Fats, Proteins, Vitamins, Minerals, Water

   b) Salt, Sugar, Fats, Proteins

   c) Water, Fats, Proteins, Salt

   d) Vitamins, Minerals, Fats, Carbohydrates

2. Which vitamin is essential for good eyesight?

   a) Vitamin A

   b) Vitamin C

   c) Vitamin D

   d) Vitamin E

3. Which of these is a rich source of protein?

   a) Rice

   b) Butter

   c) Eggs

   d) Bread

4. What is the primary source of energy for the human body?

   a) Proteins

   b) Carbohydrates

   c) Fats

   d) Vitamins

5. Which of the following materials is the best conductor of electricity?

a) Copper

b) Wood

c) Rubber

d) Plastic

6. What is the main ingredient in making glass?

a) Clay

b) Sand

c) Limestone

d) Iron

7. Which of the following materials is biodegradable?

a) Plastic

b) Glass

c) Paper

d) Metal

8. What is the chemical symbol for iron?

a) I

b) Fe

c) Ir

d) In

9. What is the process by which plants make their own food?

a) Respiration

b) Transpiration

c) Photosynthesis

d) Germination

10. Which part of the plant is responsible for photosynthesis?

a) Roots

b) Stem

c) Leaves

d) Flowers

11. What type of reproduction involves only one parent?

a) Sexual reproduction

b) Asexual reproduction

c) Crosspollination

d) Self-pollination

12. Which of the following is an example of a nonrenewable resource?

a) Solar energy

b) Wind energy

c) Coal

d) Hydropower

13. What is the main function of the human heart?

a) To digest food

b) To circulate blood

c) To control the nervous system

d) To filter blood

14. Which organ in the human body is responsible for removing waste?

a) Heart

b) Lungs

c) Kidneys

d) Liver

15. What is the smallest unit of life?

a) Tissue

b) Organ

c) Cell

d) Molecule

16. Which system in the human body is responsible for transporting nutrients and oxygen?

a) Digestive system

b) Respiratory system

c) Circulatory system

d) Nervous system

17. What is the process of breaking down food into smaller molecules called?

a) Respiration

b) Digestion

c) Circulation

d) Excretion

18. Which part of the plant helps in the absorption of water and minerals?

a) Stem

b) Leaves

c) Roots

d) Flowers

19. What is the main function of the stem in a plant?

a) To absorb sunlight

b) To support the plant

c) To absorb water

d) To store food

20. What is the force that pulls objects toward the center of the Earth?

a) Magnetic force

b) Frictional force

c) Gravitational force

d) Electromagnetic force

21. What is the unit of force in the SI system?

a) Joule

b) Newton

c) Watt

d) Ampere

22. Which of the following describes motion?

a) A change in position over time

b) The speed of an object

c) The direction of an object

d) The force applied to an object

23. What type of motion is exhibited by a pendulum?

a) Circular motion

b) Rectilinear motion

c) Oscillatory motion

d) Random motion

24. What is the formula to calculate speed?

a) Speed = Distance × Time

b) Speed = Distance ÷ Time

c) Speed = Time ÷ Distance

d) Speed = Time × Distance

25. What is the basic principle of a lever?

a) A small force applied over a large distance

b) A large force applied over a small distance

c) A small force applied over a small distance

d) A large force applied over a large distance

26. Which simple machine is used to lift heavy objects using less force?

a) Pulley

b) Inclined plane

c) Wheel and axle

d) Lever

27. What type of energy does a stretched rubber band have?

a) Kinetic energy

b) Potential energy

c) Thermal energy

d) Chemical energy

28. What is the role of an electric circuit?

a) To convert electricity into light

b) To control the flow of electricity

c) To store electrical energy

d) To convert electricity into sound

29. What is the function of a fuse in an electrical circuit?

a) To generate electricity

b) To store electrical energy

c) To break the circuit in case of excess current

d) To amplify the current

30. Which of the following is an example of a renewable source of energy?

a) Coal

b) Petroleum

c) Wind

d) Natural gas

31. What happens to light when it passes through a prism?

a) It is reflected

b) It is refracted and dispersed

c) It is absorbed

d) It is scattered

32. Which phenomenon is responsible for the formation of a rainbow?

a) Reflection

b) Refraction

c) Dispersion

d) Absorption

33. What causes an object to float in water?

a) Gravity

b) Density

c) Buoyancy

d) Pressure

34. What is the main cause of tides in the oceans?

a) Wind

b) The Moon's gravitational pull

c) Earth's rotation

d) Earth's magnetic field

35. What is the main reason for the changing seasons on Earth?

a) Earth's distance from the sun

b) The tilt of Earth's axis

c) Earth's rotation

d) The shape of Earth's orbit

36. What is the phenomenon of day and night caused by?

a) The Earth's rotation on its axis

b) The Earth's revolution around the sun

c) The Moon's orbit around Earth

d) The tilt of Earth's axis

37. Which of the following is a natural satellite of the Earth?

a) Mars

b) The Moon

c) The Sun

d) Venus

38. What type of resource is water considered?

a) Renewable

b) Nonrenewable

c) Fossil fuel

d) Mineral

39. What is the process of recycling used for?

a) To create new products from old materials

b) To burn waste materials

c) To bury waste in landfills

d) To remove impurities from materials

40. What is the most abundant gas in Earth's atmosphere?

a) Oxygen

b) Carbon dioxide

c) Nitrogen

d) Hydrogen

41. What is the primary source of freshwater on Earth?

a) Oceans

b) Rivers and lakes

c) Glaciers

d) Underground water

42. What is the main cause of water pollution?

a) Natural disasters

b) Industrial waste and sewage

c) Erosion of soil

d) Evaporation

43. Which resource is considered a fossil fuel?

a) Solar energy

b) Wind energy

c) Coal

d) Water

44. What is the main purpose of the ozone layer?

a) To provide oxygen to the atmosphere

b) To absorb harmful UV radiation from the sun

c) To regulate Earth's temperature

d) To block meteors from entering Earth's atmosphere

45. Which of the following materials is used in the construction of solar panels?

a) Silicon

b) Aluminum

c) Copper

d) Iron

46. What is the main component of natural gas?

a) Methane

b) Ethane

c) Propane

d) Butane

47. What is the process by which plants lose water vapor through their leaves?

a) Photosynthesis

b) Transpiration

c) Respiration

d) Germination

48. Which of the following is an example of a non-biodegradable material?

a) Paper

b) Glass

c) Cotton

d) Vegetable peels

49. What is the primary source of energy for all living organisms?

a) Water

b) Sunlight

c) Oxygen

d) Soil

50. Which type of simple machine is a seesaw?

a) Inclined plane

b) Lever

c) Pulley

d) Wheel and axle

Questions: Part 2

51. What is the chemical formula for water?

a) H_2O

b) CO_2

c) O_2

d) H_2SO_4

52. Which of the following organisms is a producer in a food chain?

a) Lion

b) Deer

c) Grass

d) Human

53. What is the primary gas responsible for global warming?

a) Oxygen

b) Nitrogen

c) Carbon dioxide

d) Hydrogen

54. Which of these is a secondary consumer in a food chain?

a) Rabbit

b) Grass

c) Lion

d) Grasshopper

55. What is the process by which solid waste is converted into useful products?

a) Decomposition

b) Recycling

c) Composting

d) Incineration

56. What is the primary function of red blood cells?

a) Fight infections

b) Carry oxygen

c) Clot blood

d) Produce hormones

57. What do we call the process by which a liquid changes into a gas?

a) Melting

b) Freezing

c) Condensation

d) Evaporation

58. What is the main function of the human liver?

a) Pump blood

b) Filter toxins from the blood

c) Absorb nutrients

d) Store vitamins

59. What kind of energy is stored in a battery?

a) Kinetic energy

b) Thermal energy

c) Chemical energy

d) Potential energy

60. What is the process by which green plants produce their food?

a) Respiration

b) Transpiration

c) Photosynthesis

d) Decomposition

61. Which of the following is an example of a herbivore?

a) Lion

b) Wolf

c) Deer

d) Tiger

62. Which natural phenomenon is responsible for the phases of the Moon?

a) Rotation of the Earth

b) Revolution of the Moon around the Earth

c) Rotation of the Moon

d) Revolution of the Earth around the Sun

63. Which of the following is not a function of the human skeleton?

a) Support the body

b) Protect internal organs

c) Produce blood cells

d) Absorb nutrients

64. What is the primary function of the kidneys in the human body?

a) Filter waste from the blood

b) Digest food

c) Produce energy

d) Circulate blood

65. Which of the following is a renewable resource?

a) Coal

b) Oil

c) Wind

d) Natural gas

66. What type of energy is produced by wind turbines?

a) Chemical energy

b) Electrical energy

c) Thermal energy

d) Mechanical energy

67. Which of the following is the main source of energy for the water cycle?

a) Wind

b) The Moon

c) The Sun

d) Earth's rotation

68. Which type of soil is best for growing crops?

a) Sandy soil

b) Clay soil

c) Loamy soil

d) Gravel

69. What is the chemical formula for carbon dioxide?

a) CO

b) CO_2

c) C_2O

d) C_2O_2

70. Which natural resource is used to make paper?

a) Sand

b) Clay

c) Wood

d) Iron

71. What is the force that opposes motion between two surfaces in contact?

a) Gravitational force

b) Frictional force

c) Magnetic force

d) Electrical force

72. What is the basic unit of all living organisms?

a) Atom

b) Molecule

c) Cell

d) Organ

73. Which of the following is a simple machine used to raise a flag on a flagpole?

a) Pulley

b) Inclined plane

c) Lever

d) Wheel and axle

74. Which of the following is a property of metals?

a) Poor conductors of electricity

b) High melting points

c) Brittle

d) Nonmalleable

75. What is the primary reason for deforestation?

a) Urbanization

b) Soil erosion

c) Natural disasters

d) Climate change

76. What type of energy is possessed by a moving car?

a) Kinetic energy

b) Potential energy

c) Thermal energy

d) Chemical energy

77. What is the term for the conversion of liquid water to water vapor?

a) Sublimation

b) Condensation

c) Evaporation

d) Precipitation

78. Which of these is not a greenhouse gas?

a) Carbon dioxide

b) Methane

c) Oxygen

d) Nitrous oxide

79. Which process in the water cycle is responsible for the formation of clouds?

a) Evaporation

b) Condensation

c) Precipitation

d) Transpiration

80. Which of the following is an example of a physical change?

a) Rusting of iron

b) Burning of wood

c) Melting of ice

d) Cooking of food

81. Which gas is most abundant in Earth's atmosphere?

a) Oxygen

b) Nitrogen

c) Carbon dioxide

d) Hydrogen

82. Which part of the cell contains genetic material?

a) Nucleus

b) Mitochondria

c) Cytoplasm

d) Cell membrane

83. What is the main function of the human respiratory system?

a) To digest food

b) To circulate blood

c) To exchange gases

d) To produce hormones

84. Which planet is known as the "Red Planet"?

a) Mercury

b) Venus

c) Mars

d) Jupiter

85. What is the process by which plants convert sunlight into food?

a) Respiration

b) Photosynthesis

c) Transpiration

d) Germination

86. Which natural disaster is caused by the sudden movement of the Earth's crust?

a) Tornado

b) Earthquake

c) Tsunami

d) Hurricane

87. What is the primary purpose of the roots of a plant?

a) To absorb sunlight

b) To absorb water and nutrients

c) To produce food

d) To reproduce

88. What is the term for the variety of life found in a particular ecosystem?

a) Ecology

b) Habitat

c) Biodiversity

d) Conservation

89. Which of the following is an example of a vertebrate?

a) Earthworm

b) Starfish

c) Fish

d) Jellyfish

90. What is the function of chlorophyll in plants?

a) Absorb water

b) Absorb sunlight

c) Absorb nutrients

d) Absorb oxygen

91. Which type of energy is associated with objects in motion?

a) Potential energy

b) Chemical energy

c) Thermal energy

d) Kinetic energy

92. Which of the following is a consequence of deforestation?

a) Increased oxygen levels

b) Soil erosion

c) Decreased carbon dioxide levels

d) Stabilized climate

93. What is the primary component of Earth's inner core?

a) Iron and nickel

b) Carbon and oxygen

c) Silicon and aluminum

d) Magnesium and calcium

94. Which type of natural resource can be replenished over time?

a) Non-renewable resource

b) Fossil fuel

c) Renewable resource

d) Mineral resource

95. What process is responsible for breaking down rocks into smaller pieces over time?

a) Erosion

b) Weathering

c) Deposition

d) Sedimentation

96. What is the term for the movement of water through the soil?

a) Infiltration

b) Evaporation

c) Condensation

d) Precipitation

97. What causes the Earth's seasons?

a) The Earth's rotation

b) The Earth's distance from the Sun

c) The tilt of the Earth's axis

d) The gravitational pull of the Moon

98. Which type of cloud is typically associated with thunderstorms?

a) Cirrus

b) Stratus

c) Cumulus

d) Cumulonimbus

99. What is the primary gas found in natural gas?

a) Butane

b) Propane

c) Ethane

d) Methane

100. Which device is used to measure atmospheric pressure?

a) Thermometer

b) Barometer

c) Hygrometer

d) Anemometer

Questions – Part 3

101. Which gas is released during photosynthesis?

a) Carbon dioxide

b) Oxygen

c) Nitrogen

d) Hydrogen

102. Which is the main ore of aluminium?

a) Hematite

b) Bauxite

c) Galena

d) Magnetite

103. Which part of the human brain controls balance and posture?

a) Cerebrum

b) Cerebellum

c) Medulla

d) Pons

104. Which type of reproduction involves fusion of gametes?

a) Sexual reproduction

b) Asexual reproduction

c) Budding

d) Binary fission

105. Which device converts chemical energy into electrical energy?

a) Solar cell

b) Electric cell

c) Transformer

d) Generator

106. Which metal is the best conductor of electricity?

a) Aluminium

b) Copper

c) Silver

d) Iron

107. Which part of the plant carries water and minerals?

a) Phloem

b) Xylem

c) Cambium

d) Stomata

108. What is the SI unit of power?

a) Joule

b) Watt

c) Newton

d) Ampere

109. Which gas is known as marsh gas?

a) Carbon dioxide

b) Methane

c) Ethane

d) Propane

110. Which organ in humans produces insulin?

a) Liver

b) Pancreas

c) Kidney

d) Stomach

111. Which part of the cell contains genetic material?

a) Nucleus

b) Cytoplasm

c) Cell membrane

d) Mitochondria

112. Which energy transformation takes place in an electric bulb?

a) Electrical to chemical

b) Electrical to light and heat

c) Light to electrical

d) Chemical to electrical

113. Which planet is called the “Blue Planet”?

a) Earth

b) Neptune

c) Uranus

d) Venus

114. Which part of the plant develops into seeds?

a) Ovule

b) Ovary

c) Stigma

d) Style

115. Which natural resource is used in making pencils?

a) Graphite

b) Coal

c) Charcoal

d) Lead

116. What is the process of purifying water by boiling and cooling called?

a) Sedimentation

b) Filtration

c) Distillation

d) Chlorination

117. Which of the following is a non-renewable resource?

a) Solar energy

b) Wind energy

c) Petroleum

d) Water

118. Which vitamin helps in blood clotting?

a) Vitamin A

b) Vitamin C

c) Vitamin D

d) Vitamin K

119. Which type of mirror is used in a dentist's clinic?

a) Plane mirror

b) Concave mirror

c) Convex mirror

d) None of these

120. Which blood group is known as the universal donor?

a) A

b) B

c) AB

d) O

121. Which part of the human respiratory system contains vocal cords?

a) Lungs

b) Trachea

c) Larynx

d) Bronchi

122. Which gas is used in the process of photosynthesis?

a) Oxygen

b) Carbon dioxide

c) Nitrogen

d) Hydrogen

123. What is the main function of white blood cells?

a) Carry oxygen

b) Fight infections

c) Help in clotting

d) Carry nutrients

124. Which is the largest bone in the human body?

a) Tibia

b) Femur

c) Humerus

d) Fibula

125. Which is the main component of biogas?

a) Carbon dioxide

b) Methane

c) Oxygen

d) Nitrogen

126. Which type of energy is stored in a stretched rubber band?

a) Kinetic energy

b) Potential energy

c) Heat energy

d) Sound energy

127. Which planet is known for having a storm called the “Great Red Spot”?

a) Saturn

b) Jupiter

c) Neptune

d) Uranus

128. Which organ in the human body is responsible for filtering blood?

a) Heart

b) Kidney

c) Liver

d) Lungs

129. Which is the most abundant gas in the Earth's atmosphere?

a) Oxygen

b) Nitrogen

c) Carbon dioxide

d) Argon

130. Which instrument is used to measure atmospheric pressure?

a) Thermometer

b) Barometer

c) Anemometer

d) Hygrometer

131. Which acid is found in vinegar?

a) Citric acid

b) Acetic acid

c) Lactic acid

d) Tartaric acid

132. Which organelle is called the powerhouse of the cell?

a) Nucleus

b) Ribosome

c) Mitochondria

d) Chloroplast

133. Which gas is used for the preservation of food items?

a) Oxygen

b) Carbon dioxide

c) Nitrogen

d) Hydrogen

134. Which part of the eye is responsible for focusing light?

a) Retina

b) Iris

c) Lens

d) Cornea

135. Which process changes water into ice?

a) Condensation

b) Freezing

c) Evaporation

d) Melting

136. Which is the smallest planet in the solar system?

a) Mars

b) Mercury

c) Pluto

d) Venus

137. Which part of the plant carries prepared food?

a) Xylem

b) Phloem

c) Roots

d) Flowers

138. Which force pulls objects towards the Earth?

a) Magnetic force

b) Gravity

c) Friction

d) Electric force

139. Which vitamin is produced when skin is exposed to sunlight?

a) Vitamin A

b) Vitamin C

c) Vitamin D

d) Vitamin E

140. Which is the outermost layer of the Earth?

a) Core

b) Mantle

c) Crust

d) Lithosphere

141. Which process is used to separate cream from milk?

a) Filtration

b) Centrifugation

c) Decantation

d) Sedimentation

142. Which blood cells carry oxygen?

a) White blood cells

b) Red blood cells

c) Platelets

d) Plasma

143. Which gas is responsible for the greenhouse effect?

a) Nitrogen

b) Carbon dioxide

c) Oxygen

d) Argon

144. Which type of lens is used to correct short-sightedness?

a) Convex lens

b) Concave lens

c) Cylindrical lens

d) Plano lens

145. Which element is essential for the formation of hemoglobin?

a) Calcium

b) Iron

c) Magnesium

d) Sodium

146. Which organ removes nitrogenous waste from the blood?

a) Lungs

b) Liver

c) Kidneys

d) Stomach

147. Which planet is called the “Morning Star”?

a) Venus

b) Mercury

c) Mars

d) Jupiter

148. Which part of the ear helps in hearing?

a) Cochlea

b) Eardrum

c) Semicircular canals

d) Pinna

149. Which gas is used in fire extinguishers?

a) Oxygen

b) Carbon dioxide

c) Nitrogen

d) Hydrogen

150. Which type of energy is produced by moving water?

a) Solar energy

b) Wind energy

c) Hydroelectric energy

d) Geothermal energy

Answers

1. a) Carbohydrates, Fats, Proteins, Vitamins, Minerals, Water

2. a) Vitamin A

3. c) Eggs

4. b) Carbohydrates

5. a) Copper

6. b) Sand

7. c) Paper

8. b) Fe

9. c) Photosynthesis

10. c) Leaves

11. b) Asexual reproduction

12. c) Coal

13. b) To circulate blood

14. c) Kidneys

15. c) Cell

16. c) Circulatory system

17. b) Digestion

18. c) Roots

19. b) To support the plant

20. c) Gravitational force

21. b) Newton

22. a) A change in position over time

23. c) Oscillatory motion

24. b) Speed = Distance ÷ Time

25. a) A small force applied over a large distance

26. a) Pulley

27. b) Potential energy

28. b) To control the flow of electricity

29. c) To break the circuit in case of excess current

30. c) Wind

31. b) It is refracted and dispersed

32. c) Dispersion

33. c) Buoyancy

34. b) The Moon's gravitational pull

35. b) The tilt of Earth's axis

36. a) The Earth's rotation on its axis

37. b) The Moon

38. a) Renewable

39. a) To create new products from old materials

40. c) Nitrogen

41. b) Rivers and lakes

42. b) Industrial waste and sewage

43. c) Coal

44. b) To absorb harmful UV radiation from the sun

45. a) Silicon

46. a) Methane

47. b) Transpiration

48. b) Glass

49. b) Sunlight

50. b) Lever

51. a) H_2O

52. c) Grass

53. c) Carbon dioxide

54. c) Lion

55. b) Recycling

56. b) Carry oxygen

57. d) Evaporation

58. b) Filter toxins from the blood

59. c) Chemical energy

60. c) Photosynthesis

61. c) Deer

62. b) Revolution of the Moon around the Earth

63. d) Absorb nutrients

64. a) Filter waste from the blood

65. c) Wind

66. b) Electrical energy

67. c) The Sun

68. c) Loamy soil

69. b) CO_2

70. c) Wood

71. b) Frictional force

72. c) Cell

73. a) Pulley

74. b) High melting points

75. a) Urbanization

76. a) Kinetic energy

77. c) Evaporation

78. c) Oxygen

79. b) Condensation

80. c) Melting of ice

81. b) Nitrogen

82. a) Nucleus

83. c) To exchange gases

84. c) Mars

85. b) Photosynthesis

86. b) Earthquake

87. b) To absorb water and nutrients

88. c) Biodiversity

89. c) Fish

90. b) Absorb sunlight

91. d) Kinetic energy

92. b) Soil erosion

93. a) Iron and nickel

94. c) Renewable resource

95. b) Weathering

96. a) Infiltration

97. c) The tilt of the Earth's axis

98. d) Cumulonimbus

99. d) Methane

100. b) Barometer

101. b) Oxygen

102. b) Bauxite

103. b) Cerebellum

104. a) Sexual reproduction

105. b) Electric cell

106. c) Silver

107. b) Xylem

108. b) Watt

109. b) Methane

110. b) Pancreas

111. a) Nucleus

112. b) Electrical to light and heat

113. a) Earth

114. a) Ovule

115. a) Graphite

116. c) Distillation

117. c) Petroleum

118. d) Vitamin K

119. b) Concave mirror

120. d) O

121. c) Larynx

122. b) Carbon dioxide

123. b) Fight infections

124. b) Femur

125. b) Methane

126. b) Potential energy

127. b) Jupiter

128. b) Kidney

129. b) Nitrogen

130. b) Barometer

131. b) Acetic acid

132. c) Mitochondria

133. c) Nitrogen

134. c) Lens

135. b) Freezing

136. b) Mercury

137. b) Phloem

138. b) Gravity

139. c) Vitamin D

140. c) Crust

141. b) Centrifugation

142. b) Red blood cells

143. b) Carbon dioxide

144. b) Concave lens

145. b) Iron

146. c) Kidneys

147. a) Venus

148. a) Cochlea

149. b) Carbon dioxide
150. c) Hydroelectric energy